# THE POETRY OF CERIUM

# The Poetry of Cerium

Walter the Educator

**SKB**

Silent King Books a WhichHead Imprint

Copyright © 2023 by Walter the Educator

All rights reserved. No part of this book may be reproduced in any manner whatsoever without written permission except in the case of brief quotations embodied in critical articles and reviews.

First Printing, 2023

Disclaimer
This book is a literary work; poems are not about specific persons, locations, situations, and/or circumstances unless mentioned in a historical context. This book is for entertainment and informational purposes only. The author and publisher offer this information without warranties expressed or implied. No matter the grounds, neither the author nor the publisher will be accountable for any losses, injuries, or other damages caused by the reader's use of this book. The use of this book acknowledges an understanding and acceptance of this disclaimer.

"Earning a degree in chemistry changed my life!"
- Walter the Educator

dedicated to all the chemistry lovers, like myself, across the world

# CONTENTS

Dedication . . . . . . . . . . . . . . v

Why I Created This Book? . . . . . . . . 1

**One** - Oh, Cerium . . . . . . . . . . . 2

**Two** - Celestial Jewel . . . . . . . . . 4

**Three** - Cosmic Dream . . . . . . . . . 6

**Four** - Mysterious And Pure . . . . . . 8

**Five** - Gem . . . . . . . . . . . . . . 10

**Six** - Radiant . . . . . . . . . . . . . 12

**Seven** - Enigmatic Dream . . . . . . . 14

**Eight** - Steadfast . . . . . . . . . . . 16

**Nine** - Breaking Every Chain . . . . . 18

**Ten** - Secure . . . . . . . . . . . . . 20

**Eleven** - Masterpiece Untold . . . . . 22

**Twelve** - Partner . . . . . . . . . . . 24

**Thirteen** - Forever Shines . . . . . 26

**Fourteen** - Bonfire . . . . . 28

**Fifteen** - Nature's Hand . . . . . 30

**Sixteen** - Day After Day . . . . . 32

**Seventeen** - Inspiring Harmony . . . . . 34

**Eighteen** - Bursting At The Seams . . . . . 36

**Nineteen** - Eternity . . . . . 38

**Twenty** - So Fine . . . . . 40

**Twenty-One** - Cerium's Legacy . . . . . 42

**Twenty-Two** - Honor This Element . . . . . 44

**Twenty-Three** - Divine And Sublime . . . . . 46

**Twenty-Four** - Celestial Luminary . . . . . 48

**Twenty-Five** - Magician . . . . . 50

**Twenty-Six** - Through The Night . . . . . 52

**Twenty-Seven** - Forever Will Glow . . . . . 54

**Twenty-Eight** - In Your Presence . . . . . 56

**Twenty-Nine** - Embodiment Of Love . . . . . 58

**Thirty** - The Conductor . . . . . 60

**Thirty-One** - Cherish This Element . . . . . 62

**Thirty-Two** - Dances In The Cosmos . . . . . 64

**Thirty-Three** - Grand And True . . . . . . . 66

**Thirty-Four** - Profound And Unique . . . . . 68

About The Author . . . . . . . . . . . . . . 70

# WHY I CREATED THIS BOOK?

Creating a poetry book about the chemical element Cerium was a unique and intriguing endeavor. Cerium, with its atomic number 58, belongs to the lanthanide series and has various interesting properties. By exploring the characteristics and symbolism of Cerium through poetry, I can delve into themes such as transformation, resilience, and the beauty of imperfection. Additionally, this project can be an opportunity to merge science and art, bridging the gap between two seemingly different disciplines and inspiring a deeper appreciation for the wonders of the natural world.

# ONE

# OH, CERIUM

In the realm of elements, a gem lies,
Cerium, a marvel that catches the eyes.
A rare treasure, with a spark so bright,
A dance of electrons, a celestial light.
    A symbol of power, strength, and might,
Cerium's presence, a celestial delight.
With atomic number fifty-eight,
Its secrets, the universe did create.
    In nature's crucible, it did take form,
A beacon of hope, amidst the storm.
A metal so malleable, yet strong,
Cerium's melody, a celestial song.
    Within its core, a fire burns,
An inner flame that brightly yearns.

To ignite the world, with its vibrant blaze,
Cerium's energy, an eternal phase.
    A guardian of Earth, it does embrace,
With its power, it mends and leaves no trace.
In catalysis, it plays its part,
A catalyst of change, a work of art.
    Oh, Cerium, element of grace,
Your luminescence, we can't erase.
A testament to nature's grand design,
Cerium, a jewel that will forever shine.

# TWO

# CELESTIAL JEWEL

Within the realm of elements, behold Cerium's reign,
A luminary in the periodic domain.
A silent guardian, hidden in plain sight,
Cerium's essence, a clandestine light.
    A phoenix rising from embers of the past,
Its atomic number, a secret steadfast.
With neodymium, a celestial dance,
Cerium's harmony, a cosmic romance.
    In the depths of Earth, where secrets reside,
Cerium emerges, an enigma to confide.
A chameleon in nature, it adapts with grace,
Transforming its properties, leaving no trace.
    A catalyst of change, a conductor of fire,
Cerium's alchemy, a divine choir.

Unyielding in strength, yet gentle in its might,
It breathes life into darkness, igniting the night.

From glass to steel, Cerium lends a hand,
Enhancing our world with its wizardry grand.
A guardian of engines, a savior of skies,
Cerium's touch, an engineer's prize.

Oh, Cerium, element of wonder and allure,
In your mystique, we find a treasure pure.
A paradox of power and subtlety,
Cerium, a testament to chemistry's decree.

So let us celebrate this element rare,
With reverence and awe, let us share.
For Cerium's legacy, forever shall remain,
A celestial jewel in the alchemist's domain.

# THREE

## COSMIC DREAM

    In the realm of elements, Cerium takes its stance,
A luminescent marvel, a celestial dance.
With atomic grace, it captivates the night,
Cerium's radiance, a celestial light.
    Born from the stars, in the chasms of space,
Cerium brings harmony, a celestial embrace.
Its atomic number, a secret it keeps,
In its core, a world of wonders peeps.
    A spectrum of colors, it paints the sky,
Cerium's luminescence, a celestial dye.
From fiery reds to ethereal greens,
Its vibrant hues, a celestial scene.
    A guardian of time, it never wearies,
Cerium's resilience, a celestial theory.
In atomic clocks, it counts the seconds,
A testament to precision, its essence beckons.

A catalyst of change, it sparks the flame,
Cerium's magic, a celestial game.
From catalytic converters to glass production,
Its transformative touch, a cosmic construction.

Oh, Cerium, element of awe and might,
In your presence, the universe takes flight.
A celestial symphony, you compose,
Cerium, a cosmic marvel, nobody knows.

So let us celebrate this element divine,
With reverence and wonder, let us align.
For Cerium's legacy, forever shall gleam,
A celestial enchantment, a cosmic dream.

# FOUR

## MYSTERIOUS AND PURE

In the tapestry of elements, Cerium does reside,
A whisper in the cosmos, a celestial guide.
With atomic grace, it weaves a cosmic spell,
Cerium's enchantment, a celestial tale to tell.
    A hidden gem, a luminary in the night,
Cerium's radiance, a celestial delight.
Its secrets lie within, a mystery untold,
In its core, a universe begins to unfold.
    A conductor of fire, it dances with the flame,
Cerium's alchemy, a celestial acclaim.
From lanterns to fireworks, it ignites the sky,
A symphony of light, a celestial lullaby.
    A guardian of Earth, it nurtures and protects,
Cerium's presence, a celestial aspect.

In magnets and alloys, it lends its strength,
A force of nature, a celestial wavelength.
    Oh, Cerium, element of wonder and grace,
In your essence, the universe finds its place.
A cosmic alchemist, transforming the mundane,
Cerium, a celestial gift we cannot explain.
    So let us honor this element profound,
With reverence and awe, let its essence surround.
For Cerium's legacy, forever shall endure,
A celestial treasure, mysterious and pure.

# FIVE

# GEM

Cerium, a celestial jewel of the Earth,
A chemical element of tremendous worth.
With atomic number 58, it stands,
In the periodic table, its place grand.
    A metal so rare and versatile,
Cerium's beauty knows no guile.
Its lustrous sheen, a sight to behold,
A radiant glow, precious like gold.
    From the depths of the Earth, it springs,
Amidst the rocks and mineral rings.
A guardian of nature, it lends a hand,
With healing properties, a divine command.
    Cerium, the magician of oxidation,
In catalysts, it brings transformation.

From car engines to industrial might,
It reduces pollution, a shining light.

But Cerium's wonders don't end there,
In glass manufacturing, it plays its share.
With its heat-resistant properties profound,
Cerium glass, a treasure renowned.

Oh, Cerium, element of hidden might,
In your presence, the world shines bright.
A cosmic alchemist, with secrets untold,
Cerium, a gift for the curious and bold.

Let us celebrate this element rare,
With wonder and gratitude, let us share.
For Cerium's legacy, forever shall remain,
A celestial gem, a treasure we sustain.

# SIX

# RADIANT

In the realm of elements, Cerium resides,
A symphony of atoms, where magic abides.
With a heavenly glow, a celestial sheen,
Cerium, the enchantress of the chemical scene.
   In the depths of the Earth, it patiently waits,
A guardian of secrets, the alchemical gates.
With a touch of its essence, transformations occur,
Cerium, the alchemist, a cosmic connoisseur.
   From lanterns to screens, it brings forth its light,
A luminescent dance in the darkest of night.
Cerium, the illuminator, casting a glow,
Guiding us forward in the paths we must go.
   In alloys it strengthens, with a resilient embrace,
Cerium, the protector, in the metal's embrace.

A shield against time, a guardian it stands,
Preserving the beauty of crafted hands.
    Oh, Cerium, element of wonder and grace,
Your presence, a celestial embrace.
A cosmic companion, both gentle and strong,
Cerium, the muse of this elemental song.
    Let us celebrate this element divine,
With reverence and awe, let our voices align.
For Cerium's legacy, forever shall endure,
A celestial gift, radiant and pure.

# SEVEN

# ENIGMATIC DREAM

Silent guardian of the periodic realm,
Cerium, a gem in nature's helm.
With a luminous touch, it graces the land,
A celestial conductor, an alchemist's hand.
In the glass, it captures the light,
Cerium, a prism, painting colors so bright.
From amber to ruby, a kaleidoscope dance,
A symphony of hues, a cosmic romance.
With each electron, a secret it unveils,
Cerium, a puzzle that forever prevails.
In fluorescent bulbs, it brings the day,
A celestial mimic, chasing darkness away.
Oh, Cerium, element of secrets untold,
In your presence, mysteries unfold.
A cosmic enigma, both humble and grand,
Cerium, a riddle in nature's command.

Let us celebrate this element divine,
With curiosity and wonder, let us intertwine.
For Cerium's legacy, forever shall gleam,
A celestial treasure, an enigmatic dream.

# EIGHT

# STEADFAST

Cerium, the celestial conductor of change,
In its core, a cosmic power arranged.
From fire to ice, it molds with its might,
A shape-shifting marvel, a celestial light.
   In self-cleaning ovens, it works its charm,
Cerium, the purifier, keeping things warm.
A guardian of purity, a celestial guard,
Removing impurities with a touch so hard.
   Oh, Cerium, element of alchemical grace,
In your essence, the universe finds its place.
A cosmic chemist, creating wonders anew,
Cerium, a celestial secret we pursue.
   Let us honor this element with utmost respect,
Its versatility and strength, we must reflect.

For Cerium's legacy, forever shall endure,
A celestial treasure, steadfast and pure.

# NINE

# BREAKING EVERY CHAIN

Cerium, a symphony in the earthly realm,
An element of wonder, at the helm.
From the depths of the Earth, it does emerge,
A celestial gift, its secrets it will urge.
    In flints and lighters, it sparks the flame,
Igniting passions, fueling life's game.
A catalyst of change, it takes its role,
A cosmic alchemist, transforming the whole.
    Oh, Cerium, element of transformation,
In your presence, we find inspiration.
A celestial sculptor, shaping our fate,
Cerium, the catalyst of the great.
    Let us honor this element of might,
With gratitude and awe, we hold it tight.

For Cerium's legacy, forever will remain,
A celestial force, breaking every chain.

# TEN

## SECURE

Cerium, the harbinger of dawn's first light,
A luminary in the celestial height.
In matches struck, it ignites the flame,
A catalyst of fire, a celestial game.
    A guardian of glass, it lends its might,
In lenses and prisms, bending the light.
A cosmic artist, painting rainbows above,
Cerium, the magician, inspiring with love.
    Oh, Cerium, element of enchantment and grace,
In your presence, the universe finds its place.
A celestial alchemist, weaving magic so rare,
Cerium, the enchanter, casting dreams in the air.
    Let us celebrate this element divine,
With wonder and awe, let our spirits entwine.

For Cerium's legacy, forever shall endure,
A celestial treasure, pure and secure.

# ELEVEN

# MASTERPIECE UNTOLD

Cerium, the guardian of twilight's embrace,
A shimmering presence, adorning the space.
In phosphors it dances, a celestial show,
Illuminating the darkness, casting a glow.
    A sorcerer of time, it holds the key,
In catalytic converters, it sets us free.
A cosmic engineer, purifying the air,
Cerium, the visionary, with expertise rare.
    Oh, Cerium, element of balance and grace,
In your essence, the universe finds its place.
A celestial harmonizer, orchestrating the flow,
Cerium, the conductor, making chaos slow.
    Let us honor this element with utmost respect,
Its versatility and strength, we must reflect.

For Cerium's legacy, forever shall endure,
A celestial treasure, steadfast and pure.

In nature's grand symphony, Cerium plays,
A cosmic musician, in celestial arrays.
With each atom, a melody it sings,
Cerium, the composer, creating beautiful things.

Oh, Cerium, element of artistic delight,
In your presence, the universe takes flight.
A celestial creator, crafting wonders anew,
Cerium, a cosmic muse we pursue.

Let us celebrate this element divine,
With awe and inspiration, let our spirits align.
For Cerium's legacy, forever shall unfold,
A celestial gift, a masterpiece untold.

# TWELVE

# PARTNER

Cerium, a gem in nature's embrace,
A celestial marvel, full of grace.
In the soil, it dwells with humble might,
A cosmic presence, shimmering in light.
    A guardian of health, it heals the land,
In fertilizers, it lends a helping hand.
A nurturing force, a celestial guide,
Cerium, the cultivator, by Mother Earth's side.
    Oh, Cerium, element of growth and green,
In your presence, life's wonders are seen.
A cosmic gardener, tending to the Earth,
Cerium, a celestial blessing of rebirth.
    Let us honor this element's worth,
With gratitude and care, for its nurturing birth.

For Cerium's legacy, forever shall thrive,
A celestial partner, keeping life alive.

# THIRTEEN

## FOREVER SHINES

In the depths of the Earth, a treasure lies,
Cerium, the element that mesmerizes.
With atomic grace, it shines so bright,
A celestial gem, a beacon of light.
    In catalytic converters, it purifies,
Removing toxins, it safeguards the skies.
A cosmic warrior, fighting pollution's might,
Cerium, the guardian, bringing clarity to sight.
    Oh, Cerium, element of purity and might,
In your presence, the world finds respite.
A celestial healer, mending Earth's wounds,
Cerium, a savior, in nature's interludes.
    Let us celebrate this element's embrace,
With reverence and awe, let our spirits trace.

For Cerium's legacy, forever it shall endure,
A celestial force, noble and pure.
    In our quest for progress, let us remember,
Cerium's wisdom, a celestial ember.
For in balance and harmony, we shall find,
A world where Cerium's light forever shines.

# FOURTEEN

# BONFIRE

Cerium, a hidden gem of the elements,
In the Earth's embrace, its beauty presents.
A cosmic enigma, mysterious and rare,
Cerium, the wonder beyond compare.
    In catalysts, it dances with precision,
A cosmic conductor, orchestrating with vision.
A catalyst of dreams, it sparks the fire,
Cerium, the muse, igniting desire.
    Oh, Cerium, element of dreams and hope,
In your presence, possibilities elope.
A celestial architect, shaping the unknown,
Cerium, the creator, in the seeds it's sown.
    Let us celebrate this element's allure,
With admiration and wonder, let our spirits endure.

For Cerium's legacy, forever shall inspire,
A celestial spark, a cosmic bonfire.

# FIFTEEN

# NATURE'S HAND

In the depths of the Earth, a hidden gem,
Cerium, the element that does transcend.
With a touch of magic, it comes alive,
A cosmic dancer, ready to thrive.
    In lanterns and screens, it casts its glow,
A celestial light, that continues to grow.
A guiding beacon, in the darkest night,
Cerium, the luminary, shining so bright.
    Oh, Cerium, element of illumination,
In your presence, we find fascination.
A celestial artist, painting the skies,
Cerium, the brushstroke, that mesmerizes.
    Let us celebrate this element divine,
With awe and wonder, let our spirits align.

For Cerium's legacy, forever will stand,
A celestial gift, crafted by nature's hand.

# SIXTEEN

# DAY AFTER DAY

In the depths of Earth, a treasure lies,
Cerium, the element that mystifies.
A cosmic alchemist, it weaves its spell,
Transforming secrets, only time can tell.
    With strength and resilience, it stands tall,
A celestial warrior, protecting us all.
In magnets and alloys, it finds its place,
Cerium, the guardian, embracing space.
    Oh, Cerium, element of fortitude and might,
In your presence, darkness turns to light.
A celestial shield, defending the weak,
Cerium, the sentinel, steadfast and meek.
    Let us celebrate this element divine,
With reverence and gratitude, let our souls intertwine.

For Cerium's legacy, forever it shall endure,
A celestial force, noble and pure.

In the tapestry of elements, it shines,
A cosmic jewel, with powers so fine.
Cerium, the luminary, guiding our way,
A celestial companion, day after day.

# SEVENTEEN

# INSPIRING HARMONY

In the depths of the earth, a treasure untold,
Cerium, the element, with secrets it holds.
A cosmic alchemist, transforming the unknown,
Cerium, the sorcerer, on nature's throne.
   With a touch of sorcery, it alters its form,
A celestial magician, defying the norm.
Cerium's allure, a mesmerizing spell,
In molecules and compounds, it weaves so well.
   Oh, Cerium, element of enchantment and grace,
In your presence, the world finds solace.
A celestial enchanter, weaving dreams anew,
Cerium, the catalyst, creating what's true.
   Let us celebrate this element's charm,
With wonder and awe, let our spirits disarm.

For Cerium's legacy, forever shall shine,
A celestial gift, a treasure divine.
   In the fabric of existence, it leaves its mark,
A cosmic weaver, stitching light in the dark.
Cerium, the maestro, orchestrating the symphony,
A celestial conductor, inspiring harmony.

# EIGHTEEN

# BURSTING AT THE SEAMS

In the depths of Earth, a treasure untold,
Cerium, the element, radiant and bold.
With atomic might, it dances unseen,
A celestial secret, a cosmic routine.
   In glass and ceramics, it lends its strength,
Cerium, the artisan, crafting at length.
A creator of beauty, a master of form,
Cerium, the sculptor, in the alchemist's swarm.
   Oh, Cerium, element of transformation,
In your presence, there's endless creation.
A celestial alchemist, changing the game,
Cerium, the catalyst, igniting the flame.
   Let us honor this element's hidden power,
With wonder and reverence, each passing hour.

For Cerium's legacy, forever will inspire,
A celestial muse, a cosmic fire.

In the tapestry of elements, it weaves,
A cosmic architect, creating with ease.
Cerium, the magician, conjuring dreams,
A celestial force, bursting at the seams.

# NINETEEN

# ETERNITY

In the realm of elements, a gem so rare,
Cerium, the mystic, beyond compare.
With its secrets veiled, it captivates the mind,
A cosmic enigma, forever undefined.
  Oh, Cerium, element of intrigue and awe,
In your presence, we wander without a flaw.
A celestial riddle, unraveling the unknown,
Cerium, the enchanter, on its cosmic throne.
  Let us celebrate this element's mystic allure,
With wonder and curiosity, let our spirits endure.
For Cerium's legacy, forever shall inspire,
A celestial spark, a cosmic fire.
  In the dance of particles, it takes its stance,
A cosmic conductor, orchestrating chance.

Cerium, the maestro, composing the symphony,
A celestial melody, resonating in harmony.
   Oh, Cerium, element of the enigmatic,
In your presence, the universe becomes cinematic.
A celestial storyteller, weaving tales untold,
Cerium, the luminary, shining bold.
   Let us explore this element's cosmic embrace,
With awe and reverence, let our spirits interlace.
For Cerium's legacy, forever it shall be,
A celestial wonder, for all eternity.

# TWENTY

# SO FINE

In the realm of elements, a secret lies,
Cerium, the enigma that mystifies.
A cosmic chameleon, it changes its hue,
A celestial puzzle, waiting to pursue.
   Oh, Cerium, element of intrigue and disguise,
In your presence, reality defies.
A celestial magician, shifting the scene,
Cerium, the illusionist, captivating and keen.
   Let us unravel this element's hidden tale,
With curiosity and wonder, let us prevail.
For Cerium's legacy, forever will unfold,
A celestial mystery, waiting to be told.
   In the fabric of nature, it finds its place,
A cosmic weaver, spinning cosmic lace.
Cerium, the artist, painting the skies,
A celestial masterpiece, before our eyes.

Oh, Cerium, element of mystery and grace,
In your presence, the universe finds its space.
A celestial traveler, exploring the unknown,
Cerium, the wanderer, in realms yet to be shown.
Let us celebrate this element divine,
With reverence and awe, let our spirits intertwine.
For Cerium's legacy, forever will shine,
A celestial gift, a treasure so fine.

# TWENTY-ONE

# CERIUM'S LEGACY

Cerium, the alchemist's delight,
A mystical element, shining so bright.
In the heart of the forge, it dances and gleams,
A celestial flame, igniting dreams.
Oh, Cerium, element of transformation,
In your presence, magic finds its formation.
A celestial magician, bending reality's rules,
Cerium, the sorcerer, enchanting the fools.
Let us marvel at this element's might,
With wonder and awe, in its celestial light.
For Cerium's legacy, forever shall inspire,
A cosmic enchanter, setting souls on fire.
In the tapestry of elements, it weaves,
A cosmic weaver, creating new beliefs.
Cerium, the craftsman, forging destiny's mold,
A celestial blacksmith, turning lead into gold.

Oh, Cerium, element of infinite grace,
In your presence, miracles take place.
A celestial guide, leading us to the unknown,
Cerium, the compass, our celestial throne.

Let us celebrate this element divine,
With reverence and joy, in every design.
For Cerium's legacy, forever will be,
A celestial gift, a cosmic symphony.

# TWENTY-TWO

# HONOR THIS ELEMENT

In the realm of elements, a jewel shines bright,
Cerium, the luminary, casting celestial light.
A cosmic beacon, guiding our way,
A celestial ember, igniting the fray.
   Oh, Cerium, element of endless allure,
In your presence, the universe finds a cure.
A celestial healer, mending broken hearts,
Cerium, the alchemist, where magic imparts.
   Let us revel in this element's divine embrace,
With wonder and gratitude, let our spirits interlace.
For Cerium's legacy, forever will glow,
A celestial legacy, for all to bestow.
   In the symphony of atoms, it plays its part,
A cosmic composer, crafting an intricate chart.

Cerium, the maestro, orchestrating the dance,
A celestial muse, inspiring cosmic romance.
    Oh, Cerium, element of infinite grace,
In your presence, the cosmos finds its place.
A celestial luminary, shining through the night,
Cerium, the eternal flame, forever burning bright.
    Let us honor this element's celestial might,
With reverence and awe, in its ethereal light.
For Cerium's legacy, forever will endure,
A cosmic treasure, mysterious and pure.

# TWENTY-THREE

# DIVINE AND SUBLIME

In the realm of elements, a gem so rare,
Cerium, the wanderer, with secrets to share.
A celestial traveler, traversing the unknown,
Cerium, the explorer, in realms yet unblown.
Oh, Cerium, element of cosmic allure,
In your presence, the universe finds a cure.
A celestial healer, mending broken stars,
Cerium, the alchemist, removing cosmic scars.
Let us celebrate this element's cosmic grace,
With wonder and awe, let our spirits embrace.
For Cerium's legacy, forever transcends,
A cosmic catalyst, on which creation depends.
In the symphony of atoms, it plays its tune,
A celestial conductor, orchestrating the moon.

Cerium, the magician, transforming the night,
A cosmic magician, with powers of light.

    Oh, Cerium, element of magic and charm,
In your presence, the universe finds no harm.
A celestial guardian, protecting the skies,
Cerium, the sentinel, with celestial eyes.

    Let us honor this element's cosmic might,
With reverence and awe, in its celestial light.
For Cerium's legacy, forever will shine,
A cosmic treasure, divine and sublime.

# TWENTY-FOUR

## CELESTIAL LUMINARY

In the realm of elements, Cerium takes its place,
A cosmic dancer, adorned with celestial grace.
With its luminescent glow, it paints the night sky,
A celestial artist, unveiling beauty so high.
   Oh, Cerium, element of radiant allure,
In your presence, the cosmos feels pure.
A celestial beacon, guiding lost hearts,
Cerium, the compass, leading to new starts.
   In the cosmic tapestry, it weaves a tale,
A celestial storyteller, whispering secrets unveiled.
Cerium, the enigma, shrouded in mystery,
A cosmic riddle, inviting minds to query.
   Let us celebrate this element divine,
With wonder and awe, in every design.

For Cerium's legacy, forever will endure,
A celestial gift, a cosmic treasure so pure.

In the cosmic symphony, it plays a tune,
A celestial conductor, harmonizing the moon.
Cerium, the maestro, orchestrating the night,
A cosmic composer, filling the world with delight.

Oh, Cerium, element of cosmic delight,
In your presence, the universe shines bright.
A celestial luminary, guiding souls afar,
Cerium, the celestial, our guiding star.

# TWENTY-FIVE

# MAGICIAN

In the realm of elements, a celestial jewel,
Cerium, the enchanter, captivating and cool.
A cosmic alchemist, shifting its state,
From solid to liquid, it metamorphoses with grace.
    Oh, Cerium, the chameleon of the stars,
In your essence, the universe unbars.
A celestial dancer, twirling with grace,
Cerium, the performer, captivating every space.
    With its luminescent glow, it lights up the night,
A cosmic lantern, casting a celestial light.
Cerium, the illuminator, guiding us through,
A celestial beacon, showing paths anew.
    Let us honor this element, so unique,
With reverence and awe, our hearts speak.

For Cerium's legacy, forever will shine,
A cosmic treasure, a celestial design.
　In the tapestry of elements, it weaves,
A cosmic weaver, creating cosmic beliefs.
Cerium, the artist, painting the sky,
A celestial masterpiece, captivating every eye.
　Oh, Cerium, element of wonder and might,
In your presence, the cosmos ignites.
A cosmic magician, enchanting the night,
Cerium, the celestial, casting its cosmic light.

# TWENTY-SIX

# THROUGH THE NIGHT

In the realm of elements, where secrets reside,
Cerium emerges, a cosmic guide.
A celestial traveler, roaming far and wide,
Cerium, the wanderer, with mysteries to confide.
    With its atomic dance, a celestial waltz,
Cerium enchants, as the universe exalts.
A cosmic charmer, captivating the sky,
Cerium, the enchanter, with magic in its eye.
    Oh, Cerium, element of celestial grace,
In your presence, the cosmos finds its place.
A cosmic alchemist, transforming the mundane,
Cerium, the sorcerer, bringing change in its reign.
    Let us honor this element, so profound,
With reverence and awe, our voices resound.

For Cerium's legacy, forever will endure,
A celestial gift, radiant and pure.
    In the tapestry of creation, it weaves,
A cosmic artisan, crafting dreams and beliefs.
Cerium, the creator, shaping destinies bright,
A celestial architect, sculpting wonders in sight.
    Oh, Cerium, element of cosmic art,
In your presence, inspiration sparks in every heart.
A cosmic luminary, shining with cosmic might,
Cerium, the celestial, guiding us through the night.

# TWENTY-SEVEN

# FOREVER WILL GLOW

In the realm of elements, a luminary we find,
Cerium, the enigmatic, captivating the mind.
A cosmic conductor, orchestrating the stars,
Cerium, the celestial, playing melodies from afar.
   Let us celebrate this element, divine and rare,
With reverence and awe, in the cosmic air.
For Cerium's legacy, forever will endure,
A celestial treasure, radiant and pure.
   In the symphony of creation, it takes the lead,
A cosmic composer, fulfilling every need.
Cerium, the maestro, harmonizing the spheres,
A celestial muse, inspiring cosmic cheers.
   Oh, Cerium, element of celestial might,
In your presence, the universe shines bright.

A cosmic catalyst, igniting cosmic flame,
Cerium, the eternal, forever earning acclaim.
   Let us honor this element, so grand and true,
With admiration and wonder, in all that we do.
For Cerium's legacy, forever will glow,
A celestial gift, a cosmic wonder to bestow.

# TWENTY-EIGHT

# IN YOUR PRESENCE

In the realm of elements, a gem we find,
Cerium, the jewel that captivates the mind.
A cosmic treasure, rare and sublime,
A celestial essence, transcending time.
   With a shimmering grace, it lights the way,
Cerium, the beacon that brightens the day.
A celestial guardian, guiding us through,
A cosmic protector, steadfast and true.
   Oh, Cerium, element of celestial glow,
In your presence, the universe starts to show.
A cosmic luminary, a radiant star,
Cerium, the luminescent, admired from afar.
   Let us honor this element, so divine,
With reverence and awe, for it truly shines.

For Cerium's legacy, forever will endure,
A celestial gift, enchanting and pure.
　In the symphony of creation, it plays its part,
Cerium, the conductor, igniting every heart.
A celestial maestro, orchestrating the dance,
A cosmic composer, infusing life with chance.
　Oh, Cerium, element of cosmic allure,
In your presence, the universe feels secure.
A cosmic essence, a celestial delight,
Cerium, the eternal flame, forever burning bright.

# TWENTY-NINE

# EMBODIMENT OF LOVE

In the depths of the cosmos, a shimmering hue,
Cerium emerges, a celestial debut.
A cosmic enigma, a mystery untold,
In its atomic essence, secrets to unfold.
    Oh, Cerium, element of celestial grace,
In your luminescence, the stars find their place.
A cosmic alchemist, with transformative might,
Cerium, the catalyst, igniting cosmic light.
    Let us honor this element, rare and sublime,
With reverence and wonder, throughout space and time.
For Cerium's legacy, forever will endure,
A cosmic jewel, radiant and pure.
    In the cosmic tapestry, it weaves a tale,

Cerium, the storyteller, its secrets prevail.
A celestial guide, illuminating the way,
A beacon of wisdom, through the astral array.
   Oh, Cerium, element of cosmic allure,
In your presence, the universe feels secure.
A celestial guardian, watching from above,
Cerium, the celestial, embodiment of love.

# THIRTY

## THE CONDUCTOR

Amidst the celestial symphony's embrace,
Cerium, the conductor, orchestrates with grace.
A cosmic maestro, commanding the night,
A luminary presence, casting ethereal light.
    Let us celebrate this element's reign,
With awe and reverence, we shall remain.
For Cerium's legacy, eternal and pure,
A cosmic treasure, steadfast and sure.
    In the cosmos' dance, it takes the lead,
Cerium, the luminescent, fulfilling our need.
A celestial artisan, crafting cosmic dreams,
A radiant force, as profound as it seems.
    Oh, Cerium, element of cosmic allure,
In your essence, celestial wonders endure.

A guiding star, in the vast cosmic sea,
Cerium, the celestial, forever shall be.
    Let us marvel at this element's might,
Its luminescence casting celestial light.
For Cerium's legacy, we shall adore,
A cosmic beacon, forevermore.

# THIRTY-ONE

# CHERISH THIS ELEMENT

In the realm of elements, a gemstone gleams,
Cerium, the enchanting, in cosmic dreams.
A celestial jewel, its brilliance untold,
A radiant light, a cosmic story unfolds.

Oh, Cerium, element of cosmic grace,
In your presence, the cosmos finds its place.
A celestial dancer, twirling through the night,
Cerium, the luminary, shimmering with delight.

Let us honor this element, rare and true,
With reverence and wonder, in all that we do.
For Cerium's legacy, forever will endure,
A celestial gift, captivating and pure.

In the symphony of stars, it plays its part,
Cerium, the conductor, igniting every heart.

A cosmic composer, harmonizing the spheres,
A celestial maestro, soothing cosmic fears.
   Oh, Cerium, element of cosmic delight,
In your essence, the universe ignites.
A cosmic catalyst, sparking cosmic flame,
Cerium, the eternal, forever earning acclaim.
   Let us cherish this element, celestial and bright,
For in Cerium's presence, we find cosmic light.
A celestial treasure, a guiding cosmic force,
Cerium, the celestial, our eternal source.

# THIRTY-TWO

# DANCES IN THE COSMOS

In the realm of elements, a cosmic gem appears,
Cerium, the luminary, banishing all fears.
A celestial dancer, with grace and finesse,
Cerium, the enchanter, brings cosmic caress.
    Let us celebrate this element, rare and sublime,
With awe and wonder, throughout space and time.
For Cerium's legacy, forever shall endure,
A cosmic treasure, radiant and pure.
    In the cosmic symphony, it takes center stage,
Cerium, the conductor, commanding cosmic rage.
A celestial architect, building dreams to behold,
A cosmic alchemist, turning lead into gold.
    Oh, Cerium, element of celestial might,
In your presence, the universe shines bright.

A cosmic catalyst, igniting cosmic flame,
Cerium, the eternal, forever earning acclaim.
    Let us honor this element, with reverence and grace,
As it dances in the cosmos, leaving a luminous trace.
For Cerium's legacy, forever will remain,
A celestial gift, a cosmic jewel to sustain.

# THIRTY-THREE

# GRAND AND TRUE

In the realm of elements, a gem does shine,
Cerium, the celestial jewel, so divine.
With elegance and grace, it takes its place,
A cosmic luminary, a celestial embrace.
    Oh, Cerium, element of cosmic allure,
Your presence in the universe is pure.
A celestial magician, enchanting the skies,
With your mystical powers, you mesmerize.
    In the cosmic alchemy, you hold the key,
Transforming the elements, with celestial decree.
A catalyst of change, in the celestial dance,
Cerium, the alchemist, creating cosmic chance.
    Oh, Cerium, the cosmic weaver of dreams,
In your essence, cosmic harmony gleams.

With every atom, you spark cosmic fire,
A radiant force, lifting us higher and higher.

Let us celebrate this element, so rare,
With wonder and awe, in the cosmic air.
For Cerium's legacy, forever will endure,
A celestial treasure, radiant and pure.

In the symphony of creation, it takes the lead,
A cosmic composer, fulfilling every need.
Cerium, the maestro, harmonizing the spheres,
A celestial muse, inspiring cosmic cheers.

Oh, Cerium, element of celestial might,
In your presence, the universe shines bright.
A cosmic catalyst, igniting cosmic flame,
Cerium, the eternal, forever earning acclaim.

Let us honor this element, so grand and true,
With admiration and wonder, in all that we do.
For Cerium's legacy, forever will glow,
A celestial gift, a cosmic wonder to bestow.

# THIRTY-FOUR

## PROFOUND AND UNIQUE

In the realm of elements, a celestial spark,
Cerium, the mystic, igniting the dark.
A cosmic luminary, enchanting the night,
Cerium, the celestial, a shimmering light.
   Within its atomic embrace, secrets reside,
Cerium, the enigma, where mysteries hide.
A cosmic alchemist, transforming the unknown,
Unleashing its power, a celestial throne.
   Oh, Cerium, element of cosmic birth,
In your presence, the universe finds its worth.
A cosmic conductor, orchestrating the stars,
Guiding our journeys, no matter how far.
   Let us honor this element, celestial and rare,

With reverence and awe, let's show our care.
For Cerium's legacy, forever it will gleam,
A cosmic treasure, a muse of our dreams.

In the cosmic tapestry, it weaves its design,
Cerium, the celestial, a jewel, so fine.
A cosmic catalyst, sparking cosmic flame,
Cerium, the eternal, forever earning acclaim.

So let us embrace this element, profound and unique,
In its celestial grace, let our spirits speak.
For Cerium's legacy, forever it will be,
A cosmic companion, for eternity.

# ABOUT THE AUTHOR

Walter the Educator is one of the pseudonyms for Walter Anderson. Formally educated in Chemistry, Business, and Education, he is an educator, an author, a diverse entrepreneur, and he is the son of a disabled war veteran. "Walter the Educator" shares his time between educating and creating. He holds interests and owns several creative projects that entertain, enlighten, enhance, and educate, hoping to inspire and motivate you.

Follow, find new works, and stay up to date with Walter the Educator™
at WaltertheEducator.com

www.ingramcontent.com/pod-product-compliance
Lightning Source LLC
LaVergne TN
LVHW010603070526
838199LV00063BA/5060